SCIENTIFIC AMERICAN EDUCATIONAL PUBLISHING

SCIENCE PARTY

10 FUN

EXPERIMENTS FOR STUDENTS

BRING SCIENCE HOME

Published in 2025 by The Rosen Publishing Group, Inc.
2544 Clinton Street, Buffalo, NY 14224

Contains material from Scientific American , a division of Springer Nature America, Inc., reprinted by permission as well as original material from The Rosen Publishing Group.

First Edition

Editor: Kristen Rajczak Nelson
Designer: Rachel Rising

Activity on page 5 by Svenja Lohner and Science Buddies (November 7, 2019); page 9 by Science Buddies (October 29, 2015), page 15 by Science Buddies (September 13, 2012); page 21 by Science Buddies (February 16, 2012); page 27 by Megan Arnett and Science Buddies (November 17, 2016); page 33 by Science Buddies (April 4, 2013); page 39 by Science Buddies (March 7, 2013); page 43 by Science Buddies (January 7, 2016); page 49 by Svenja Lohner and Science Buddies (February 2, 2017); page 55 by Ben Finio and Science Buddies (April 23, 2015).

All illustrations by Continuum Content Solutions

Photo Credits: pp. 4, 5, 9, 15, 21, 27, 33, 39, 43, 49, 55 Anna Frajtova/Shutterstock.com; pp. 5, 8, 9, 13, 15, 18,21, 24, 33, 37, 49, 53 cve iv/Shutterstock.com.

Library of Congress Cataloging-in-Publication Data

Names: Scientific American, inc., editor.
Title: Science party : 10 fun experiments for students / Scientific American editors.
Description: Buffalo, NY : Scientific American Educational Publishing, an imprint of Rosen Publishing, [2025] | Series: Bring science home | Includes bibliographical references and index. | Audience: Grades 7-9.
Identifiers: LCCN 2023057681 (print) | LCCN 2023057682 (ebook) | ISBN 9781725350670 (library binding) | ISBN 9781725350663 (paperback) | ISBN 9781725350687 (ebook)
Subjects: LCSH: Science--Experiments--Juvenile literature.
Classification: LCC Q182.3 .S3438 2025 (print) | LCC Q182.3 (ebook) | DDC 507.8--dc23/eng/20231226
LC record available at https://lccn.loc.gov/2023057681
LC ebook record available at https://lccn.loc.gov/2023057682

Manufactured in the United States of America

CPSIA Compliance Information: Batch #CSSA25. For further information contact Rosen Publishing, New York, New York at 1-800-237-9932.

CONTENTS

⚛ THESE ACTIVITIES INCLUDE SCIENCE FAIR PROJECT IDEAS.

INTRODUCTION

Sometimes we think of scientists as people in white coats who work in a lab, but you can be a scientist at home too! Using common household items such as balloons, vinegar, milk, food coloring, and candy, as well as the scientific method, you can try these fun experiments to learn about the world around you.

Projects marked with  include a section called Science Fair Project Ideas. These ideas can help you develop your own original science fair project. Science fair judges tend to reward creative thought and imagination, and it helps if you are really interested in your project. You will also need to follow the scientific method. See page 61 for more information about that.

Balloon Magic with Bernoulli's Principle

USE PHYSICS TO DEFY EXPECTATIONS WITH THIS BREEZY, EASY ACTIVITY.

PROJECT TIME

15 minutes

KEY CONCEPTS

Physics
Pressure
Bernoulli's principle
Airflow

Have you ever seen pictures or videos of a roof being blown off a house during a hurricane or tornado? You might be surprised to hear that the roof is actually not blown off by the strong winds but instead by the air inside the house! This can be explained by Bernoulli's principle, which states that fast-moving fluids or air, such as strong winds, have lower pressure than slow-moving air. In a hurricane, the fast flow of air above the roof's shape generates a low-pressure area. This creates a pressure difference between the air outside and the air inside the house. Eventually the pressure difference becomes large enough that the air inside the house starts pushing on the roof. The roof experiences lift, similar to an airplane wing, and flies away! In this activity, you will put Bernoulli's principle to work—but do not worry, your roof will be safe!

BACKGROUND

Daniel Bernoulli was a Swiss scientist who in the 18th century studied how fluids behave when they are in motion. When experimenting with fluids flowing through an hourglass-shaped tube, he made a discovery. He realized that fast-moving fluids produce less pressure and slow-moving fluids produce greater pressure. His discovery became known as the Bernoulli principle. It is not only true for fluids but also for air because gases—just like fluids—are able to flow and take on different shapes. A simple demonstration of Bernoulli's principle requires floating a Ping-Pong ball in a moving stream of air, for example on top of a fan or hairdryer pointed straight up. Why does the ball not fly off the fan? It is because of Bernoulli's principle. The fast-moving air, which carries the ball into the air, is at a lower pressure than the air surrounding the ball. When the ball starts to fall off the column of air above the fan, the surrounding higher-pressure air pushes the ball back into the area of lower pressure above the fan. As a result, the ball stays afloat on top of the fan.

Bernoulli's principle can also explain how lift is generated under an airplane wing. Airplane wings are designed to let the air flowing over the top move faster than the air flowing underneath. This creates a pressure difference in which the pressure on the top of the wing is lower than on the bottom. This higher-pressure air pushes up on the wing and thus creates an upward lifting force (similar to the roof example above). There are many more demonstrations of Bernoulli's principle in the real world. You are going to perform one of them in this activity. It doesn't involve flying or blowing a roof off a house, but it will be just as impressive—in fact it will almost be magical!

MATERIALS

- Two balloons of the same size
- String (about 24 inches [61 cm] long)
- Scissors
- Tape
- A door frame
- Paper towel tube

PREPARATION

- Inflate both balloons and tie them off at their ends. Both balloons should be approximately the same size.

- Cut two pieces of string, each about 12 inches (30 cm) in length.
- Tie the end of one string to one of the balloons.
- Tie the end of the other string to the second balloon.
- Use tape to attach the loose end of each of the strings to the underside of the top of the door frame. Space the balloons so that there is a gap of about 6 inches (15 cm) between them.
- Make sure to keep the balloons away from significant air flow (such as a vent or fan).

PROCEDURE

- Step in front of the balloons and hold the paper towel tube so that you can blow air into the space between the two balloons. *What do you expect to happen to the balloons if you blow air in between them?*
- Make sure that the balloons are still. Then blow into the paper towel tube very slowly. Try to produce a steady air flow. *What do you notice? Are the balloons moving?*
- If the balloons moved, stop their movement, and then blow in between them again using the paper towel tube. This time try to blow through the tube harder than before but still try to maintain a steady air flow. *What happens to the balloons this time? Can you explain your observations?*
- Repeat the previous step but this time blow through the tube as hard as you can, producing a steady air flow. *Do your results change with increasing air flow? Why or why not?*

EXTRA

Repeat the same tests but vary the size of your paper tube. *Do smaller diameter tubes, such as a straw, have the same effect?*

SCIENCE FAIR IDEA

Find out if you can still make the balloons move if they are farther apart from each other. Change the distance between the balloons, and test if it affects your results. *Do you find a maximum or minimum distance beyond which the activity does not work anymore?*

EXTRA

Instead of balloons, try using Ping-Pong balls or other objects in this activity. *Do you still see the same effect with larger or smaller or heavier objects?*

OBSERVATIONS AND RESULTS

Did you notice that both balloons magically moved toward each other without being touched at all? The effect which you observed is a great demonstration of Bernoulli's principle. As long as both balloons just dangle from the doorframe, the air around them in every direction is static. This means the air exerts the same amount of pressure onto every side of the balloon and both balloons are still. When you slowly blew air in between the balloons, they probably did not move much. This is because a very slow air flow does not greatly change the pressure conditions around the balloons.

When you blew through the paper tube more forcefully, however, you should have noticed that the balloons (almost!) magically came together. By blowing air forcefully between the balloons you created an area of low pressure. This is because fast-moving air produces less pressure. The air pressure between the balloons decreased in comparison with the air pressure around the rest of the balloons. Because higher pressure pushes toward lower pressure the balloons were pushed toward each other. You could have made the same observations using Ping-Pong balls instead of balloons. With heavier objects, however, the generated air pressure difference might not be enough to make them move. The spacing between the balloons matters as well. If the gap between the balloons is too wide, the low air pressure area produced by blowing between them will no longer have an effect.

CLEANUP

Return all materials to the places where you found them.

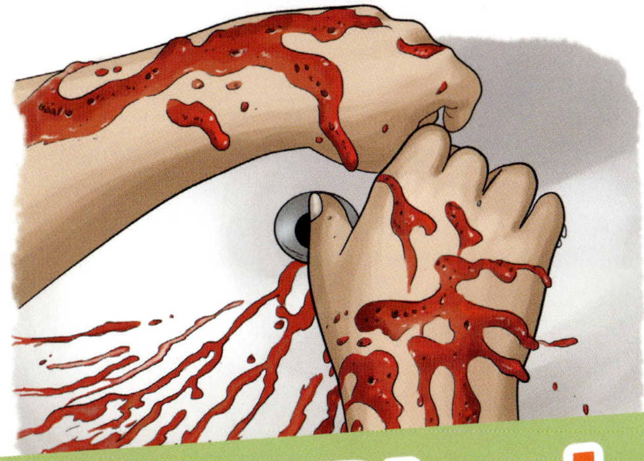

Fake Blood Made Scientific

LOOKING TO FAKE A GHASTLY GASH FOR A HALLOWEEN COSTUME? FEARSOME FANGS? LEARN HOW TO MAKE MIXTURES THAT LOOK (AND BEHAVE) LIKE REAL BLOOD—AND ARE ALL EDIBLE!

Whether it is dripping vampire teeth or a leaky bandage, fake blood is sure to bring Halloween characters to life—so to speak. Although chocolate syrup might make convincing fake blood in black-and-white pictures, it is not so believable in full color—or in real-life encounters. In this activity, science helps you engineer good-looking (and tasty!) fake blood.

PROJECT TIME

30 minutes

KEY CONCEPTS

Physics
Viscosity
Blood

BACKGROUND

Several physical quantities, including the viscosity and color, determine the look of a fluid. Viscosity is a measure of the thickness and stickiness of a fluid. It quantifies its resistance to flow. Together with its density, the viscosity of blood determines how blood flows through blood vessels—or out of your finger after a paper cut. At normal body temperature (98.6°F [37°C]), blood is four times more viscous than water but only slightly denser. As a result, it flows more slowly, and it is stickier than water. Corn syrup or honey, on the other hand, are far more viscous (more than thousand times) than blood. They are also denser. Diluting these with water will reduce their viscosity.

To get the color of human blood right seems easy—it is always red when we see it. The red color comes from red blood cells, which carry oxygen from the lungs to tissues and organs. They turn bright red when exposed to oxygen. Clotting (or coagulating) blood has a different look. The chemical reactions taking place that clot blood make it look darker. It is also thicker and more gel-like.

Now that you know a little about what blood looks like, you are better equipped to create fake blood.

MATERIALS

- Three small cups or glasses (more if you want to test more recipes)
- Flour
- Water
- Corn syrup
- Chocolate syrup
- Red and blue liquid food coloring (Use it carefully, food coloring can stain. Work on a protected surface and clean up any accidental spills right away.)
- Cutting board or plate (preferably white or a light color)
- Three small spoons (teaspoon-size ones work well)
- Measuring spoons
- Sink or bowl with water to rinse cups and the cutting board
- Towel to dry off the cutting board
- Protected area that will not be stained by food coloring

- Old clothing that can be stained by food coloring
- Paper and pencil (optional)
- Other ingredients to try, such as ketchup, coffee grounds, maple or pancake syrup, cornstarch (optional)

PREPARATION

- Protect your workspace and clothes—the fake blood can be sticky and difficult to wash off.
- Gather your ingredients at your workspace.

PROCEDURE

- Mix 1 tablespoon of flour into 2 tablespoons (30 ml) of water in one cup. (You can keep track of all of your recipes with a pencil and paper if you like. This will allow you to easily recreate your best fake blood again!)
- Measure 3 tablespoons (45 ml) of corn syrup into a second cup.
- Measure 3 tablespoons (45 ml) of water into a third cup.
- Add two drops of red food coloring to each cup and watch how the food coloring disperses over the fluids. *Does the red food coloring disperse in the same way in all three liquids?*
- Use small spoons to mix the red food coloring with the other ingredients in all three cups. Be sure to use a dedicated spoon with each cup.
- Use the small spoons to take 1/4 teaspoon (1 ml) of red liquid from each cup and place the liquids as separate blobs in a row near the edge of a cutting board or plate. Lift the side of the surface with the blobs so the board is tilted. Watch how the liquids run down. *Do all three run down with the same speed? Do some leave a trace behind? Which liquid do you think would make the best fake blood?*

- You probably noticed that water runs way too fast to mimic blood well. Discard the contents of this cup. While you are at the sink, rinse off the cutting board and wipe it dry so it is ready to perform more tests.

- Reuse the now-empty third cup and in it, pour half of the contents from each of your other two cups. (If you'd rather start over for the third cup, here is the recipe: 1/2 tablespoons of flour, 1 tablespoon [15 ml] of water, 1 1/2 tablespoons [22 ml] of corn syrup, all mixed together with two drops of red food coloring. If you are keeping track of your recipe, note what you have added.)

- Perform your test again, placing small amounts of the three liquids on top of the cutting board or plate, tilting the board and watching how the liquids run down. *Do any of these mimic how blood might run? Which one would you say comes closest?*

- If you found that one recipe mimicked blood better than the others, continue with that recipe. If you did not find a reasonable match, proceed with using the third cup with the flour, corn syrup, water, and food coloring. You can still adjust how your fake blood flows later in the activity.

- In addition to flowing like real blood, its color needs to mimic the color of real blood. Human blood is red. Oxygen-rich blood, like the blood from a finger cut or a bloody nose, is bright red. Blood turns dark red when it has delivered its oxygen to tissue and organs. Blood drawn by a nurse has this color. Note that the blood vessels you see through your skin only appear blue; this is because light reflecting off them is filtered by your skin before it reaches your eye. If you opened up your skin, the blood vessels and the blood inside them would look red. *Which type of blood would you like to create?*

- To get a realistic look, add a tiny bit of blue food coloring and/or a little chocolate syrup. Note that the blue food coloring is very strong. It is best to make a drop of blue food coloring on your cutting board or plate, dip the tip of your mixing spoon in it, and use this spoon to stir the tiny amount of liquid in your cup. It is better to add gradually, because once you have too much blue, it is hard to correct. A tiny bit of blue or brown will mimic the bright, oxygen-rich blood. Add a little more for the darker blood that is low in oxygen.

- Place a blob on the top of your cutting board or plate, tilt the board and watch it flow. *Does it look realistic? Are the flow and the color right?* To improve your fake blood, add corn syrup, flour, water, food coloring (red or blue), or chocolate syrup—one at a time. Check the results after each addition and, if you are noting your recipe, do not forget to keep track of what you added.
- As a last test, try your blood on your skin, teeth, or other surface where the food coloring will not stain (or can stain, such as on a costume you are preparing for Halloween). *Does it look like real blood?*

EXTRA

To mimic dried blood, add more chocolate syrup to make it thicker and darker—clotting blood has a darker red color with a brown hue and is thicker, more gel-like. You can also add a spoon of coffee grounds to mimic the scaly appearance of the crust on healing wounds.

SCIENCE FAIR IDEA

Test other ingredients you find around the kitchen for their ability to mimic blood. See how they flow and how they can add to the color. Some ideas include ketchup, red beet juice, maple syrup, and peanut butter.

OBSERVATIONS AND RESULTS

The main ingredients of your final recipe probably consisted of corn syrup diluted with water and thickened with flour. This particular mixture resembles the flow of blood quite well because it has a similar viscosity, or resistance to flow. You probably noticed the red-colored corn syrup is too viscous, it flows too slowly and leaves a thick trace behind. The red water was way too runny and left almost no trace on the board. The flour helped to thicken the mixture but did not get the viscosity quite right.

The recipe to obtain the most realistic color depends on the exact hue of food coloring used. Most often, a tiny bit of blue yields a more realistic look. The brown from chocolate syrup often provides a good addition in color, especially if you want to mimic blood that is low in oxygen. It also reduces the transparency of the fake blood.

Now that you have some experience, you might just find more great recipes using other food items available in your kitchen. Enjoy your realistic (and edible) creation this Halloween!

CLEANUP

Wash all equipment with soapy water. Return materials to the places where you found them.

Sculpted Science

Turning Milk into Plastic!

TURN A TYPICAL BREAKFAST DRINK INTO PLASTIC IN THIS FUN EXPERIMENT!

Have you ever heard that plastic can be made out of milk? If this sounds far-fetched, you may be surprised to learn that from the early 1900s until about 1945, milk was commonly used to make many different plastic ornaments. This included buttons, decorative buckles, beads and other jewelry, fountain pens, the backings for handheld mirrors, and fancy comb and brush sets. Milk plastic (usually called casein plastic) was even used to make jewelry for Queen Mary of England! In this activity, you will make your own casein plastic out of hot milk and vinegar.

PROJECT TIME

25 minutes

KEY CONCEPTS

Plastic
Polymers
Milk
Casein

BACKGROUND

Plastics are a group of materials that may look or feel different but can all be molded into varied shapes. The similarities and differences between different plastic products come down to the molecules that comprise them. All plastics are composed of molecules that repeat themselves in a chain, called a polymer. Polymers can be chains of either one type of molecule or different ones, which are linked together in a regular pattern. Also, in a polymer, a single repeat of the pattern of molecules is called a monomer, which can consist of just one type of molecule or include several different kinds.

Milk contains many molecules of a protein called casein. Each casein molecule is a monomer, and a chain of casein monomers is a polymer. The polymer can be scooped up and molded, which is why plastic made from milk is called casein plastic.

MATERIALS

- Measuring cup
- Milk
- Stove-top oven and pan or a microwave and microwaveable container
- Mug or other heat-resistant cup
- Measuring spoons
- White vinegar
- Paper towels
- A clean, hard surface that will not be damaged by dampness
- Spoon
- Cookie-cutters, glitter, food coloring, markers (all optional)
- Adult help and supervision with heating and handling hot liquids

PREPARATION

- Heat 1 cup (237 ml) of milk in a pan or on a stove top until it is steaming hot. Alternatively, you can microwave the milk in a microwaveable container by warming it at 50 percent power for five minutes. It should be about the same temperature as milk you would use to make hot cocoa; heat longer if needed.

PROCEDURE

- Add 4 teaspoons (20 ml) of white vinegar to a mug or other heat-resistant cup.
- Add the cup of hot milk to the mug. You should see the milk form white clumps that are called curds. *Why do you think the milk forms curds when it is added to the vinegar? What do you think they are made of?*
- Mix the mug slowly with a spoon for a few seconds. *What happens when the milk and vinegar are mixed together? Why do you think this is?*
- Stack four layers of paper towels on a hard surface that will not be damaged if it gets damp.
- Once the milk and vinegar mixture has cooled a bit, use a spoon to scoop out the curds. You can do this by tilting the spoon against the inside of the mug to let excess liquid drain out while retaining the curds in the spoon. Collect as many curds as you can in this way and put them on top of the paper towel stack.
- Fold the edges of the paper towel stack over the curds and press down on them to absorb excess liquid. Use extra paper towels if needed to soak up the remaining moisture.
- Knead all of the curds together into a ball, as if it were dough. What you have in your hands is casein plastic. *How do the kneaded curds feel and look different from the original ones?*
- If you want to use the casein plastic to make something, you can color, shape, or mold it now (within an hour of making the plastic dough) and leave it to dry on paper towels for at least 48 hours. Once it has dried, the casein plastic will be hard.
- **Tip:** To shape the plastic, the dough must be thoroughly kneaded. Molds and cookie-cutters work well, or, with more patience, the dough can be hand sculpted. Food coloring, glitter, or other decorative bits can be added to the wet casein plastic dough, and dried casein plastic can be painted or colored with markers.

EXTRA

How does the amount of vinegar used affect the yield of casein plastic? To find out, you can repeat this activity. In addition to testing 4 teaspoons (20 ml) of white vinegar with 1 cup (237 ml) of hot milk, try testing 1 teaspoon (5 ml), 2 teaspoons (10 ml), or 8 teaspoons (40 ml) of the vinegar, each with 1 cup (237 ml) of hot milk. To collect the most curds and get a better idea of their casein plastic yield, instead of scooping the curds with a spoon, you can pour the vinegar and milk mixture through a piece of cotton cloth (such as an old T-shirt) secured with rubber bands on top of a cup.

EXTRA

In addition to vinegar, there are a lot of other acids that we encounter in the kitchen all the time, such as lemon juice, orange juice, soda pop, and tomato juice. *Do some of these common acids work better than others for making casein plastic?*

SCIENCE FAIR IDEA

You used hot milk in this activity that was not at a specific temperature, but using hotter or colder milk might affect the casein plastic reaction. Design an experiment to investigate this. *How does the temperature of the milk affect how much casein plastic you can produce?*

OBSERVATIONS AND RESULTS

Were you able to see the curds when the hot milk and vinegar were mixed together, and could you scoop out the curds to make casein plastic?

After you add the hot milk to the vinegar, small, white chunks—or curds—should become visible in the mixture. This is because adding an acid (such as vinegar) to the milk changes the latter's pH (acidity) and makes the casein molecules unfold and reorganize into a long chain, thereby curdling the milk. You should be able to use a spoon to separate the curds from most of the liquid. Additional drying of the curds with the paper towels should make the curds ready to knead into a ball and use as casein plastic, which can be molded and decorated.

CLEANUP

All materials should be washed with soapy water and returned to where you found them.

Starch Power

Generating Gooey Gels

MAKE SOME SLIMY AND USEFUL GEL IN THIS SIMPLE EXPERIMENT.

PROJECT TIME

45 minutes

KEY CONCEPTS

Chemistry
Gels
Molecules
Solutions

Have you ever wondered how gels are made? You probably have several kinds of products around your house—some that you eat—that use gels: puddings, diapers, shoe insoles, packaging, ice cream, toothpaste, and many more. A gel is a mixture of solid particles suspended in liquid. The solid particles in the gel can absorb water, causing the gel to swell and increase in volume. If you ever dunked a diaper in a tub of water, you have seen this in action; the diaper will swell as it absorbs the water, and if you cut the diaper open you will see the pieces of gel that are absorbing the water. These properties of gels make them quite useful for a variety of purposes. What are some of the things you can think of that gels can help do?

BACKGROUND

Gels are very common because they have so many uses—not only around the house but also in science. Around the house you might also find gels in knee pads, JELL-O, bike seats, Sterno fuel, air fresheners, hair products, cold packs, toys, and medicine gel capsules. In the medical field, bio-gels—gels that contain medicines—are being studied for use in the body for repair or to deliver medicines. Gels, specifically aerogels, have been used on space probes to collect star dust and return it to Earth for study. Gels have even been designed to clean up radioactive contamination.

A simple gel can be made using cornstarch and water, and varying the ratio will change its properties. Cornstarch, as its name implies, is starch derived from corn. Starch is a commonly consumed carbohydrate that is made of many bonded molecules of glucose (a type of sugar). It is made by most plants for use as an energy source. When water is added and it is heated, some of the bonds between the glucose molecules break and the starch undergoes "gelatinization," which means, you have yourself your very own, homemade gel!

MATERIALS

- Cornstarch
- Water
- Measuring spoons and cups
- Two stir sticks or forks
- Two microwave-safe bowls
- Microwave oven
- Oven mitts or small towel
- Wax paper
- A flat, hard surface (such as a kitchen countertop)
- Scratch piece of paper and pen (optional)
- Graph paper (optional)

PREPARATION

- To one bowl, add 1/3 cup (79 ml) and 1 tablespoon (15 ml) of water. To the same bowl add 1 teaspoon of cornstarch. This bowl now contains 5 percent cornstarch.

- To the second bowl, add 1/3 cup (79 ml) of water. To the same bowl add 4 teaspoons of cornstarch. This bowl contains 20 percent cornstarch.
- Each bowl should have the same total volume. Keep track of which bowl is which. You might want to make small labels using a piece of paper and a pen.
- Use extreme caution when microwaving the gels. As you heat the gel it can become superheated and cause a burn.

PROCEDURE

- Stir each bowl thoroughly (using a different stir stick for each bowl). *How do the 5 percent and 20 percent cornstarch solutions look and feel different from, or similar to, each other?*
- Heat each bowl separately in the microwave on high for a total of 1 minute and 30 seconds. Do this by heating each bowl for 30 seconds, then stopping the microwave and stirring the bowl, followed by heating for another 30 seconds, stirring again, and then heating for a final 30 seconds. Use caution when stirring the bowls and handling them as the gels can become superheated.
- Carefully remove each hot bowl from the microwave using oven mitts or a small towel. Stir the contents of each bowl after removing it. *How did the cornstarch solution in each bowl change? How does each solution look in appearance and feel when you stir it?*
- Allow the solutions to cool to room temperature before touching them as they will be very hot. It may take about 5 to 10 minutes for them to cool. *As they cool, do you notice any more changes?*
- When the solutions have cooled to room temperature, touch them with your fingers. *How does the 5 percent solution feel compared with the 20 percent solution?*

- Put a piece of wax paper on a flat, hard surface and carefully place the gel solutions, one at a time, onto the wax paper and allow each gel to spread out. If you want to quantify how much each gel spreads out, put a piece of graph paper over the wax paper before placing the gel solutions on the wax paper.
- Leave the gels undisturbed for about five minutes. *How much did the 5 percent solution spread out compared with the 20 percent solution? How else are the two solutions similar or different?*
- If you used graph paper, you can trace around the gels when they are done spreading and count the number of squares that are inside the traced area. *Which solution covered more squares? How many more squares did it cover?*

SCIENCE FAIR IDEA

Starches can come from many different sources, such as corn, potatoes, rice, and tapioca. In this activity, you used cornstarch, but you can try this activity with other kinds of starches. *Do you get similar gels using different starches?*

EXTRA

Gels can be made out of many different materials, such as gelatin, agar, diaper filling, tapioca, seaweed, or fruit pectin. You can try making gels out of these different materials. *How are gels made from different materials different? How are they similar?*

OBSERVATIONS AND RESULTS

Before microwaving, was the 20 percent cornstarch solution less opaque, whiter, and harder to stir than the 5 percent solution? After microwaving, did both solutions turn clear? Did the 20 percent solution become much thicker than the 5 percent solution? After it cooled, did the 20 percent solution spread out a lot less than the 5 percent one?

Cornstarch is made up of many molecules of glucose, specifically amylopectin and amylose. When starch is heated with water, the starch granules swell and burst, causing them to break down and release the glucose molecules into the water. Consequently, the starch molecules interact with more water, increasing the randomness of the solution. This process is known as gelatinization. When the heated solution of cornstarch and water cools down, the amylose molecules can bind each other again to create a molecular mesh. The more amylose molecules there are, the firmer, or more viscous, the mesh will be.

After heating, a solution with more starch in it, such as the 20 percent cornstarch solution compared with the 5 percent one, will be firmer and stickier. Because it is more viscous, the 20 percent solution will spread out much less than the 5 percent one once cooled. Different gels can be made using different starches because starches' consistencies vary with the proportions of amylose and amylopectin that comprise them. What could you use your gels for?

CLEANUP

Solutions with cornstarch can be very sticky and messy. The 20 percent cornstarch gel should be composted or thrown in the trash. The 5 percent solution can be thrown away or diluted with water and rinsed down the drain.

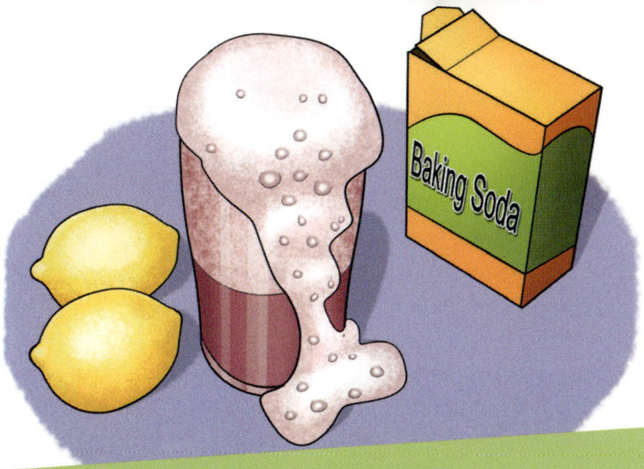

Pass the Cranberry . . . Volcano!

DID YOU KNOW THAT SOMETHING AS SIMPLE AS CRANBERRY JUICE CAN HELP YOU BE A CHEMISTRY DETECTIVE?

PROJECT TIME

25 minutes

KEY CONCEPTS

Chemistry
Acids
Bases
Reactions

You might enjoy cranberry sauce, but if you've ever tasted a real cranberry, you were probably surprised by the taste—definitely not as sweet as the sauce or sweetened dried berries! Pure cranberries and cranberry juice are very tart to eat, but they can be delicious when combined with other ingredients (such as orange zest and sugar).

Cranberries are acidic, similar to lemons or limes. Foods that are acidic will usually taste sour or tart, so we use them to brighten up a meal, but we rarely eat them by themselves. (Can you imagine eating a whole lemon—ouch!) In addition to being acidic, cranberries also contain a special, color-changing pigment that we can use to test whether something is an acid or a base. In this activity, we will use cranberries (in this case, cranberry juice) to identify acids and bases and to observe the chemical reactions created when you mix the two. Who knew that cranberries were so much more than sauce?

BACKGROUND

If you've ever gotten lemon juice in a cut and felt that sting, then you know a little bit about acids. And you probably know about bases too: When you wash your hands and get that slippery feeling from the soap—that's a base. Most liquids are either an acid or a base, but some acids and bases are stronger than others. Lemon juice stings when you get it in a cut because it's acidic. But other types of very strong acids can dissolve wood and metal. Believe it or not, the acid in your stomach (that helps dissolve your food) is a very strong acid. If you dropped a little of your stomach acid onto a block of wood, the acid would eat right through it!

To measure how acidic or basic a liquid is, we use the pH scale. This scale ranks liquids from 0 to 14, with 0 being the most acidic and 14 being the most basic. We can identify how acidic or basic something is by testing how it reacts with other things. We often use pH indicators for this purpose, which turn different colors depending on whether they're exposed to an acid or a base.

Cranberries are special because they contain a natural pH indicator—pigments called anthocyanins, which change color depending on the pH of their environment. In this activity, we will use this special ability of cranberries to observe a chemical reaction and observe how basicity or acidity changes the color of a pH indicator.

MATERIALS

- 3 cups (710 ml) of pure, 100 percent cranberry juice (not cranberry juice cocktail)
- 1 cup (237 ml) of water
- Baking soda
- Lemon juice (4 tablespoons [59 ml], approximately two lemons, as fresh-squeezed juice works best)
- Four clear glass or plastic cups
- One large plastic or glass clear bowl or casserole dish (a large glass baking pan works well)
- Measuring spoon (tablespoon)
- Measuring cup (1 cup size)
- A flat workstation that can be exposed to some spills (including cranberry juice)

PREPARATION

- Use the measuring cup to measure 1 cup (237 ml) of cranberry juice into three of the clear cups.
- Use the measuring cup again to measure 1 cup (237 ml) of water into the fourth clear cup.
- If using fresh lemon juice (recommended), juice the lemon and set the juice aside in a cup or bowl.
- Place the three cups with juice in them side by side in the big dish. We'll call the cup on the left cup 1, the middle one is cup 2, and the cup on the right is cup 3. The cup with water is the water cup, which should be set aside for now.

PROCEDURE

- Use your measuring spoon to measure 2 tablespoons (30 ml) of lemon juice into cup 1. Compare the color of the cranberry juice in cup 1, with the color of the juice in cup 2 (with no lemon). *What do you notice about the color of the liquid in cup 1? Did it change when you added the lemon juice? If so, did the lemon juice make it lighter or darker?*
- Set cup 1 aside for now.
- Using your measuring spoon, measure 1 tablespoon of baking soda and pour it into cup 3 (on the right). *What happens when the baking soda reaches the cranberry juice? What do you see? What do you hear? What happens to the juice? Is there more or less in the cup?*
- Allow the reaction to continue for at least two minutes. When there is no more foam being created, look at the juice in the bottom of cup 3. Compare it with cup 2. *What do you notice about the color of the liquid in cup 3? Is it darker, lighter or the same as that in cup 2? What about the volume of juice in cup 3? Is there more or less in cup 3 compared with cup 2? What do you think caused the changes you observe in cup 3?*

- Put cup 1 back in your dish, so that all three cups are again lined up in order. Compare the color and volume of each cup. *Which cup is the most different from the others? What makes it different?*

- Rinse off your measuring spoon so there is no baking soda on it. Then use it to add 2 tablespoons (30 ml) of lemon juice to cup 3. *What happens when you add lemon juice to cup 3? Do you notice any differences in the reaction compared with when you added baking soda? How is this reaction different? How is it the same?*

- Allow the reaction to continue for approximately two minutes, until no more foam is created. *Did anything change in cup 3 after you added the lemon juice? If so, what is different? What do you think caused the changes you observe?*

- Use your measuring spoon to measure 2 tablespoons of baking soda into the water cup. Observe what happens when you add the baking soda. *What do you notice about the water? Has anything changed? What does this tell you about the cranberry juice and baking soda reaction? Does baking soda always cause liquids to foam or only some liquids?*

EXTRA

If you have enough cranberry and lemon juice, try adding baking soda to two side-by-side cups of cranberry juice. Allow the reaction to take place. Now add lemon juice to just one of the cups. Compare the color and volume of the juice as well as the how the foam looks in the two cups. *What is different about the two cups?*

EXTRA

Try adding 2 tablespoons (30 ml) of lemon juice to a cup of cranberry juice and then adding baking soda. *How does this change the reaction?*

EXTRA

Try increasing the amount of baking soda that you add. (Make sure to keep this in your dish so it doesn't get messy!) *Does adding more baking soda make the reaction happen faster? Does it increase the amount of foam?*

EXTRA

With an adult helper, try microwaving 1 cup (237 ml) of cranberry juice for about one minute, making it warm (but not hot). Add the baking soda to the warm juice. *Does the temperature of the juice change the reaction?*

OBSERVATIONS AND RESULTS

You might have noticed that the color of the cranberry juice changed when you added lemon juice or baking soda. Adding the baking soda should have made the juice turn a darker color whereas adding the lemon juice made it slightly lighter. This takes place because of the amazing pigments that cranberries contain, the anthocyanins, which act as a pH indicator. Anthocyanins change color depending on whether they are reacting with an acid or a base. From this activity would you guess that anthocyanins get darker when they come in contact with a base or with an acid? If you predicted base, you're right! And in the presence of acids anthocyanins become lighter in color.

When you added baking soda to the cranberry juice, a reaction took place that released a gas, which creates big bubbles as well as little bubbles (foam) in the juice. When you added baking soda to the water cup, however, nothing really happened—no foam was created, just some cloudy water. Why doesn't baking soda cause a reaction in water the way it does in cranberry juice? The difference is that because cranberry juice is slightly acidic—it reacts with anything that is basic. Baking soda is a base, therefore when it comes in contact with an acid like cranberry juice, a reaction occurs. But when the liquid is pH neutral, as water is, no reaction occurs.

Finally, you might also have noticed that the volume (amount of juice) was less in cup 3 after the baking soda reaction. This is because the foam created by the reaction was made up of the juice itself. (You could probably tell because the foam was a similar color.) Therefore, there was less liquid juice left in the cup because some of the juice became foam during the reaction.

CLEANUP

Rinse cups and measuring spoons with soapy water and return your materials to where you found them. The liquids can be poured down the drain.

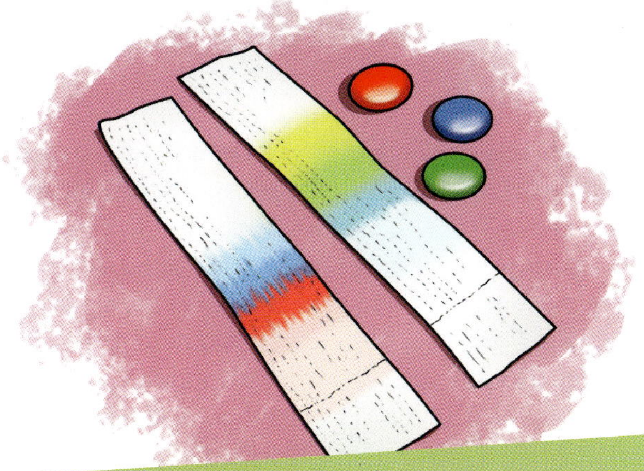

Solution Science

Colorful Candy Chromatography

THIS EXPERIMENT IS FUN AND TASTY! LEARN ABOUT THE DYES USED IN YOUR FAVORITE CANDIES.

Do you have a favorite M&M'S or Skittles candy color? Have you ever wondered what dyes are used to make that particular color? Some candy colors are the result of a single food coloring whereas others may use a combination of dyes to create just the right appearance. In this activity, you'll get to do some scientific detective work at home to investigate what different dyes are used to color M&M'S candies. This is a great activity for some leftover candies!

PROJECT TIME

30 minutes, plus drying time

KEY CONCEPTS

Chemistry
Dyes
Food science
Solubility

BACKGROUND

If you've ever seen a drop of water hit a printout from an inkjet printer or a page with marker drawings, you know that when this happens the ink starts to run. The water is absorbed into the fibers of the paper. As the water travels through and down the paper, it picks up ink particles and carries them along with it. This same process can be put to good use as a technique called paper chromatography.

Chromatography includes a group of techniques that separate the various components in a complex solution. In each chromatography setup, there is generally a so-called mobile element (a fluid in which the components are dissolved) and a stationary aspect (the material the fluid moves through). For example, in paper chromatography, water is the mobile phase and paper is the stationary phase. The solution's components ideally move at different speeds as they travel through the stationary phase so that they can be separated. In paper chromatography, different pigments can be separated based on their solubility, or their ability to dissolve in water. The more soluble a pigment is, the more readily it will dissolve in the mobile phase and the farther it will usually travel as it is carried by the water being absorbed into the paper.

MATERIALS

- Paper towels. Thicker ones will work better.
- Scissors
- Pencil
- Large bowl or pot
- Measuring spoon
- Measuring cup
- Water
- Salt
- Large jar with a narrow opening
- Medicine dropper
- Clean plate
- M&M'S, one of each color you want to test
- Timer or clock
- Red, green, and blue food coloring liquids
- Clothespins
- Drying rack (optional)

PREPARATION

- Cut the paper towels into strips that are each about 1 inch (2.5 cm) wide. Make each strip at least 1 inch (2.5 cm) longer than the height of your large jar. Cut three strips for the food coloring samples and an additional strip for each M&M color you want to test.
- Use a pencil to label which candy color or food coloring will be spotted on each paper strip.
- Draw a pencil line about 1 inch (2.5 cm) from the bottom end of each paper strip.
- In a large bowl or pot mix together 1/8 teaspoon of salt with 4 cups (1 l) of water. (To get approximately 1/8 teaspoon you can fill a 1/4 teaspoon measuring spoon so that it is half full.) Pour a small amount of the salt solution into a large jar with a narrow opening so that the jar is about 1 inch (2.5 cm) full.

PROCEDURE

- Put a single drop of water on the clean plate. Set an M&M'S candy in the drop of water and leave it there for three minutes.
- After three minutes, flip the candy over so its other side is in the drop of water. Leave it there for another three minutes. Then remove the candy.
- Take one of the paper towel strips you prepared and place the pencil line you drew on top of the now-colored drop of water. *Do you see the dye soak into the strip, coloring the area around the line?* You can get some extra dye on the line by dabbing the sides of the wet candy with it.
- Repeat these steps with the other M&M'S colors you want to test, using a new strip for each M&M'S candy.

- Next, prepare a strip for each food coloring. Put a small drop of food coloring on the plate and lightly dab the line of a new strip against it. Repeat this so that you have three food coloring strips (one each for the red, green, and blue food coloring).
- Put the pigmented end of a strip into the jar you filled with salt water. Put the strip into the jar until the strip just touches the salt water. Drape the top of the strip over the jar's opening and secure it with a clothespin. Make sure that each strip is not touching the jar's sides but only makes contact with the jar where it is secured. Secure as many of the other strips as will fit in the jar. Be sure they don't touch the jar's sides or each other.
- Let the water rise up the strips until it is about 1 to 2 inches (2.5 to 5 cm) from the top. Remove the strips at this point. (This may take around 10 minutes.) *What has happened to the colors on the strips?* Keep a close eye on the strips and where the water is—if you let the strips run too long, the water can reach the top of the strips and distort your results.
- If you could not fit your other strips in the large jar, put them in the jar now, as you did with the other strips, and run them until they are similarly done.
- Let the strips completely dry out. You can set them on a drying rack, or on the clothespins laid flat to do this.
- Look at your strips. *How many dye components does each M&M'S color have? Can you identify which dyes that are in the food coloring are also likely in the M&M'S candies? Do your results make sense to you?*

EXTRA

In this activity, you used M&M'S candies, but you could try it with other colored candies as well, such as Skittles. *Does the red in Skittles look the same as the red M&M'S when you test both using paper chromatography? What about their other colors?*

SCIENCE FAIR IDEA

A more accurate way to identify colored components in a solution using paper chromatography is by determining their retention factor (Rf value). The Rf value is the ratio between how far the component travels and the distance the mobile phase (solvent) travels from a common starting point (the line you drew on the strips). If other conditions are kept the same, the Rf value for a certain component should be consistent. You can do this activity again, but this time measure these distances and calculate the Rf value for each component. *Based on the Rf values, can you identify which dyes are used in the different colored M&M'S candies by comparing them with the known food coloring dyes?*

EXTRA

You could try this activity again, but this time compare using different kinds of mobile phases (for example, salt water, water, vegetable oil, isopropyl rubbing alcohol, etcetera). *Does a dye travel different distances depending on the mobile phase you use? What do you think this tells you about the solubility of that dye in the different mobile phases?*

OBSERVATIONS AND RESULTS

Did you find that green and brown M&M'S are each made up of two dyes whereas blue, yellow, and red M&M'S only use one dye? Did some of these dyes match the ones from the red, blue, and green food coloring strips?

You may have seen that blue food coloring liquids you tested, on the other hand, have blue and red dyes; red food coloring has two different red dyes; and green food coloring uses a blue and a yellow dye (the yellow may be covered by the blue dye, making it appear green; and all of this can depend on the exact food colorings used). Looking at the ingredients list on the packaging can help you determine exactly which government-certified Food Drug and Cosmetic Act (FD&C) dyes these are. Blue M&M'S use one of the same blue dyes and red M&M'S share one of these red dyes. (You can figure out which by looking at the ingredient list on their package.) If you tried brown M&M'S, you probably saw that they contain a red dye (the same one as red M&M'S) and a blue dye. Green M&M'S also use two dyes, a yellow dye, and a blue dye, although the latter may be too faint to easily see. Yellow M&M'S use a single yellow dye.

CLEANUP

Throw away the paper towels and wash the bowls and pots. If you want, you can eat the wet M&M'S candies!

Starchy Science

Creating Your Own Colloid

USE CORNSTARCH AND WATER TO CREATE A COLLOIDAL SOLUTION WITH SOME INTERESTING PHYSICAL PROPERTIES.

Have you ever wondered what whipped cream, jelly, and milk have in common? Aside from all being tasty, they are also all made up of tiny, solid particles that are dispersed, or distributed, in water. This type of mixture is called a colloidal solution. Colloidal solutions have some very interesting physical properties, such as acting like a solid and a liquid at the same time! In this activity, you'll get to create a colloidal solution that's made using cornstarch and water, and then explore these properties firsthand.

PROJECT TIME

20 minutes

KEY CONCEPTS

Matter
Colloidal solutions
Physical properties
States of matter

BACKGROUND

Colloidal solutions (also called colloidal suspensions) contain little particles, ranging from one to 1,000 nanometers in diameter. (A nanometer is very small—a human hair is about 100,000 nanometers wide!) In this activity, the particles used are cornstarch, and they're evenly dispersed throughout a quantity of water. These particles are so small that you can't see them with the naked eye—and when you look at the colloidal solution it appears homogenous, or uniform. If you put it under a microscope, however, it would look heterogeneous, or as if things had been mixed together.

Having the right particle size is essential for making a colloidal solution. If smaller particles are used, they will dissolve in the water and not be visible, even with a microscope. For example, think of how sugar dissolves in water, making a homogenous solution. On the other hand, if larger particles are used, such as grains of sand, they will not dissolve in the water. Instead, they separate from the water, and are so large they form a heterogeneous mixture of particles that can be seen with the naked eye. Particles of cornstarch are just the right size (about 100 to 800 nanometers in diameter) to make a colloidal solution with water.

MATERIALS

- Small bowl or cup
- Cornstarch
- Cup, mug, or drinking glass
- Water
- Medicine dropper
- Fork

PREPARATION

- Add 1 tablespoon of the cornstarch to the small bowl or cup.
- Fill the empty cup, mug, or drinking glass with water.

PROCEDURE

- Use the medicine dropper to add water from the cup to the small bowl that holds the cornstarch. Add the water one drop at a time, counting as you go. *What happens as the water touches the cornstarch?*

- After you have added 20 drops, stir the cornstarch with the fork. Break up any clumps that formed.
- Keep adding drops one at a time, stirring with the fork after every 20 drops. *What happens as you add more water?*
- Once you have added 100 drops of water total, stop for a moment and observe the cornstarch. *How does it look? How has it changed?*
- Continue adding drops, but now mix the cornstarch after every 10 drops.
- Stop adding water when all of the cornstarch has flowed together like a liquid. It will probably take around 150 to 170 drops of water total. *How many drops did it take for you? How would you describe the solution's appearance?*
- Play around with the solution in the bowl. Poke it with your finger and put some on your fork or in your hand. *How does it move? What does it feel like? What do you think this tells you about colloidal solutions and their physical properties?* **Tip:** If it seems to get chalky while you're investigating it, try adding and mixing in a few more drops of water.

EXTRA

Colloidal solutions can be made out of other common products aside from cornstarch, such as other starches (for example, potato, tapioca, and rice starches) as well as gelatin. Try making colloidal solutions out of other substances. *How do these compare with the colloidal solution made using cornstarch?*

EXTRA

Starches are often used to make gels. Try heating the solution and see what happens. *How does it change? Are there new physical properties that you can observe?*

EXTRA

Clay soil behaves like a colloidal material when it has just the right amount of water in it. You could try making a colloidal clay soil solution and test the effect of different forces on it. Colloid solutions can appear solid against strong downward forces but weak against lateral forces that push sideways. *In which direction is your colloid solution the weakest? If your sample were clay soil in the real world, how could this contribute to landslides or earthquakes?*

OBSERVATIONS AND RESULTS

Did the cornstarch solution resemble a sluggish white liquid? Did it become a hard solid when you pressed on it, but turn into a liquid when it was allowed to freely move?

Colloidal solutions can act like both a solid and a liquid. The colloidal solution you made does not behave like a typical solid or a typical liquid. Phases of matter such as liquid, gas, solid, and plasma are physical properties of matter. The cornstarch solution is a non-Newtonian fluid, which means it does not behave like a Newtonian fluid. When most liquids experience a sideways shearing force, such as by pushing your finger down against them, they move out of the way. The liquids also have a proportionate response, meaning if you push harder, they move out of the way faster. The liquids' viscosity, or resistance to flow, doesn't change. Water, and most fluids, are Newtonian and behave in this way. Because the cornstarch solution is non-Newtonian, however, you should have seen that it does not behave like this. When you put stress on it, such as by pressing down on it or poking it with the fork, it responded by becoming harder, changing to a solid state. When you did not put stress on it, for example by simply holding it in your hand, it again changed its state, this time becoming a runny liquid.

CLEANUP

Be sure to throw your cornstarch solution in the garbage, as it can clog drains. Clean materials with soapy water and return them to where you found them.

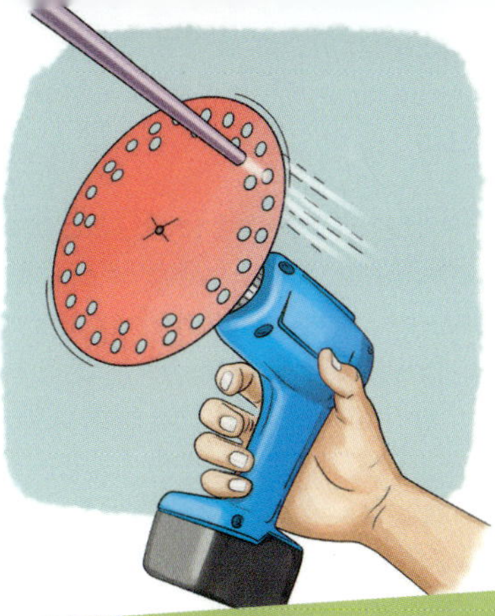

Building a Disk Siren

DID YOU KNOW THAT SIRENS CAN HELP EXPLAIN HOW SOUND FREQUENCY WORKS? AND YOU CAN MAKE YOUR VERY OWN! ROLL UP YOUR SLEEVES AND MAKE SOME NOISE TO LEARN MORE.

You probably hear them almost every day: sirens. Police cars, ambulances, fire trucks—they all can come blaring. Their wailing sound is piercingly loud and pretty effective at clearing the road in front of them. But have you ever thought about how this loud noise is generated? Make your own simple disk siren in this activity and find out for yourself.

PROJECT TIME

40 minutes

KEY CONCEPTS

Sound
Acoustics
Siren
Physics

BACKGROUND

To understand how sirens work one must first look at the physics of sound, which is basically just the movement of air. By moving or displacing air particles, you generate a sound wave, which consists of a repeating pattern of high- and low-pressure regions that travels through the air in the form of a pressure wave. The air particles vibrate in a back-and-forth motion and the rate at which this vibration occurs is called frequency.

The simplest version of a siren is a mechanical device that pushes an air stream against a rotating perforated disk. As the disk spins, the airflow will be alternately interrupted and allowed to pass. The ensuing fluctuation in air pressure produces a series of regular pressure waves that we perceive as sound. The number of holes in the disk determines its frequency. The disk's rotation speed also has an impact on the sound's tone, which is its pitch and is heard as the characteristic wailing siren sound. You can calculate the tone's precise frequency by multiplying the number of rotations per second by the number of holes in the disk.

Today electronic sirens are slowly replacing the mechanical siren design. Many of the sirens on fire trucks, however, are still based on the same principle as the disk siren. In this design, an electric motor turns a fan that spins inside a slotted drum. The fan sucks in air from the front and expels it through the sides. When the fan openings align with the slots in the drum, an airburst is generated. Similar to the perforated disk, the fan basically chops the air stream into individual airbursts and their frequency determines the siren's pitch.

That was a lot of information. Now it is time to make some noise using a self-made perforated disk siren!

MATERIALS

- Protractor
- Heavy construction paper
- Double-sided foam tape
- Scissors
- Pencil
- (Math) compass
- Round-hole puncher
- Cordless electric screwdriver (You should only use it with adult assistance.)
- Small hexagonal bit
- Straw
- Paper clip or needle
- Paper towel
- Adult helper

PREPARATION

- Draw a circle with the compass that has a 3.1 inch (8 cm) radius on the cardboard. Within the same circle draw two smaller circles with 2.8 inch (7 cm) and 2.4 inch (6 cm) radii.
- Cut out the biggest circle.
- Use the protractor to draw lines through the center of the circle in equal 15-degree intervals.
- Take the hole puncher, and on the 2.8-inch (7-cm) circle punch a hole at every intersection of the lines and the circle.
- On the innermost 2.4-inch (6-cm) circle punch holes at every second intersection of the lines and the circle.
- You should now have a disk with two concentric rings of holes in which the outermost ring has double the amount of holes compared with the innermost ring.
- Cut a small piece of double-sided foam tape and stick it on the center of one side of the cardboard disk.
- With a paper clip or needle make a small hole through the center of the circle and foam tape.
- Put a small hexagonal bit in your cordless screwdriver.
- Take the perforated cardboard disk and push the hexagonal bit through the small center hole of the circle so that the foam tape faces toward the electric screwdriver. Use the foam tape to adhere the cardboard disk to the cordless screwdriver so that it sits firmly on the screwdriver.
- Set the cordless screwdriver to its highest setting.

PROCEDURE

- Take the straw and hold your hand in front of it while blowing into it. *Can you feel the air flowing through the straw?* Listen carefully. *Does the air make any sound?*

- Ask your adult helper to hold the cordless electric screwdriver with the cardboard disk attached. Do not turn it on when anyone else is nearby. Now have your adult helper run the cordless screwdriver on the highest setting. *Is there any sound coming from the disk? Do you hear anything besides the noise of the cordless screwdriver?*

- Next, have your adult helper power the cordless screwdriver again and keep the disk spinning on the highest setting. Make sure to keep the rotation speed of the disk the same. While the disk is spinning, place the straw about 0.4 to 0.8 inch (1 to 2 cm) in front of, or perpendicular to, the outer ring of holes and blow through it. Be sure you do not let the straw touch the disk. *What happens this time? Listen again; does your siren make a sound? Do you hear a high- or low-pitch tone?*

- While the disk is still spinning at the same speed, move your straw in line with the inner circle of holes and blow air through the straw against the disk. *Can you hear a sound again? How does this tone differ from the previous one? Why do you think the sound changed?*

- Stop the cordless screwdriver, and thus the disk's rotation. Ask your adult helper to slowly power it again, but this time pay attention to the disk speed. *Is the rotation speed always the same? Can you make the disk spin at different rates?*

- Once you and your helper have figured out how to create different disk rotation speeds, use the straw again and blow against the outer ring of holes on the disk. While blowing, change the speed of the cordless screwdriver so that the disk spins at different rates. *What happens to the sound of the siren when you change the speed? How does the tone sound with a high-speed setting compared with low speed? Can you explain your observations?*

- Finally, play around with blowing at the two rings of holes at different rotation speeds. *Can you create a realistic wailing siren sound?*

EXTRA

Create more disks with different hole patterns and listen to how they sound. *What happens if the holes are not equally spaced? How does it sound if you have a series of holes followed by a gap?*

EXTRA

Take a lightweight paper, such as a paper towel, and have your adult helper hold it closely behind the spinning disk (between their body and the disk) while you blow through your straw. Observe the paper towel. *What happens to the paper towel? Does it move at all? Can you see a movement pattern?* (It helps if you spin the disk relatively slowly for this part of the activity.)

OBSERVATIONS AND RESULTS

Did you get your siren to make some wailing sounds? Just blowing through the straw on its own probably did not produce a siren sound. You only created airflow, as you probably felt with your hand, but no pressure wave. Also, if you just spin the disk with the cordless screwdriver, the only noise that you should hear is the cordless screwdriver itself. Once you push an air stream against the perforated disk with your straw, however, a sound wave is generated that you can hear. This is because the air you blow through the straw alternately puffs through the holes and is cut off as the wheel spins. If you did the extra activity with the paper towel, you probably actually saw the wave pattern created by the on-and-off airflow. You should have seen the paper towel "flapping" back and forth, depending on if the air was able to pass through the hole or not. Each of these airbursts acts like a single sound wave and many of them together create a continuous sound. The sound might not be very pure, but you probably were able to detect a smooth note.

When you blew air through the inner circle—with half the amount of the holes—you should have noticed that the sound's pitch decreased. Specifically, the tone should have been one octave lower than the tone from the outer circle, meaning that the two circles create two musical pitches with half or double frequency. The reason for this is that with the number of holes, you increase the number of created air puffs, thus increasing the sound wave's frequency. The higher a frequency, the higher a note's pitch. Thus, if you halve the number of holes, the created tone should have half the frequency. Finally, if you change the rotation speed of the disk while you are blowing against the disk, the pitch of the sound also changes. As the disk spins faster, again the puffs' frequency increases and their sound rises in pitch. By alternating the rotation speed back and forth from low to high, you can create the characteristic wailing of a siren. Every hole pattern you generate on the disk will result in a specific sound—sometimes a smooth musical tone, sometimes a rough noise. Give it a try!

CLEANUP

You can discard the cardboard disks in your recycling bin. All other materials should be put back where you found them.

A Milk-Curdling Activity

SAY "CHEESE"! LEARN HOW CHEMISTRY CAN HELP YOU TURN MILK INTO CHEESE—NO AGING REQUIRED!

PROJECT TIME

30 minutes

KEY CONCEPTS

Chemistry
Food Science
Proteins
Enzymes

Have you ever poured yourself a cup of milk and instead of a smooth liquid, all you get is clumps? This is usually a sign that the milk has gone bad. And if it smells sour, it probably has. But the physical process of what happened to the milk is called coagulation, which is the mechanism that occurs when proteins in the milk clump. Although you do not necessarily want this in your milk, without coagulation (or curdling) there would be no cheese or yogurt, which is why it is a very important process in the food industry. Wonder how you can make milk curdle—without having it be spoiled? Try this activity to find out!

BACKGROUND

Humans have turned milk into a multipurpose liquid. By itself, milk is a nutrient-rich beverage. But when you start treating milk in various processes, all kinds of other products can be created such as butter, yogurt, buttermilk, and cheese. Milk mostly consists of fat, protein, lactose (a kind of sugar), and water. The milk fat is suspended in the water as fine droplets, which makes it an emulsion. Milk also contains a lot of proteins that, in this case, are mostly whey and casein. Because casein is poorly soluble in water, its proteins build spherical structures called micelles that allow them to stay in suspension as if they were soluble.

With both fat and proteins in suspension, milk is a white liquid. The micelle structures, however, can easily be disrupted or changed, and once altered they cannot be reformed. Because the micelle holds the casein protein in suspension, without it the micelles will clump together, and the casein comes out of the solution. The result of this process of milk coagulation, or curdling, is a gelatinous material called curd.

The processes for making many other dairy products such as cottage cheese, ricotta, paneer, and cream cheese start with milk curdling. This is why cheese producers want the milk to curdle. There are different ways to start milk coagulation. You can do it with acid or heat as well as by letting the milk age long enough or with specific enzymes (which are proteins that perform a specific chemical reaction). Chymosin, for example, is an enzyme that alters the casein micelle structure to make milk curdle. Proteases are other enzymes that disrupt the casein micelle structure by chopping up proteins, causing milk to curdle. In this activity, you will try two different methods of making milk curdle—and produce some cheesy results!

MATERIALS

- Milk
- Lemon (fresh)
- Pineapple (fresh)
- Lemon squeezer
- Food grater, juicer, or blender
- Teaspoon
- Tablespoon
- Knife
- Cutting board
- Two pieces of cheesecloth or cotton fabric
- 10 small transparent, microwavable cups (that each hold about 2 ounces [59 ml])

- Paper towels
- Microwave
- Permanent marker
- Adult helper
- Workspace that can tolerate spills
- Timer (optional)

PREPARATION

- Mark four of the small cups with the labels "pineapple juice," "pineapple juice (heated)," "lemon juice," and "lemon juice (heated)."
- Take a fresh pineapple and with the help of an adult cut off the rind on a cutting board. Only use about one-fifth of the pineapple. Cut the flesh in smaller pieces and grate it. Alternatively, you can use a juicer or blender. Then, place the grated fruit in a piece of cheesecloth and squeeze at least 1 teaspoon (5 ml) of juice into each cup that you labeled with "pineapple juice."
- Put the cup that you labeled "pineapple juice (heated)" into the microwave and heat it just long enough to get it boiling (about 10 to 20 seconds). When it starts to boil, carefully take it out of the microwave and let it cool down.
- Take a fresh lemon and use the lemon squeezer to make lemon juice. Add at least 1 teaspoon (5 ml) of juice into each cup that you labeled with "lemon juice."
- Again, put the cup labeled "lemon juice (heated)" into the microwave and heat it for 10 to 20 seconds. Once it starts boiling, carefully take it out and let it cool down.
- Label four of the remaining cups "1" to "4." Fill each of these cups with about 1 tablespoon (15 ml) of milk.
- Label the last two cups "curd" and "whey."

PROCEDURE

- Place the cup that you labeled "1" in front of you. It should contain 1 tablespoon (15 ml) of milk. *How does the milk look? What happens if you gently swirl the milk in the cup? Do you notice anything unusual?*

- Use a clean teaspoon to add 1 teaspoon (5 ml) of the freshly squeezed lemon juice to your milk in cup 1. Swirl the cup slightly. *Does the milk change when you add the lemon juice? If yes, does the change occur immediately or after awhile? When you swirl the cup a little bit, what do you observe at the wall of the cup?*

- Take your second cup of milk and this time add 1 teaspoon (5 ml) of the heated lemon juice. *Do you see the same reaction happening as before? How does the milk change? Is the reaction as fast as the previous one?*

- Use your third cup of milk, and with a clean teaspoon add 1 teaspoon (5 ml) of pineapple juice to the milk. Observe what is happening for about five minutes. *Does the milk curdle with pineapple juice? Is the reaction fast or slow compared with that of the lemon juice?*

- To the fourth cup of milk add 1 teaspoon (5 ml) of heated pineapple juice and swirl the cup slightly. Again, observe the cup for about five minutes. *Do you get a similar result again or is it different? If yes, how is it different? What do you think happened? Can you explain your observations?*

- Choose the cup with milk that gave you the greatest amount of curd. Then, place your second cheesecloth over the cup labeled "whey" and carefully pour the curdled milk mixture onto the cloth. Fold the cloth over the curdled milk and squeeze the liquid from the mixture into the cup. *Do you see a lot of liquid coming out of the cheesecloth? What does the liquid look like? Is it clear, does it have a color, or does it still look like milk? What do you think the liquid is?*

- Once you have squeezed out all the liquid, open the cheesecloth and scoop the curd into the cup labeled "curd." *How much curd did you get, and what does it look like? Does it remind you of a cheese product? How does it feel if you touch it with your fingers? What parts of the milk do you think are inside the curd?*
- If you want (and only if you used clean materials!), you can taste a little bit of the whey and curd. *Does it taste similar to milk? Is it sweet, sour, creamy, or salty? Does it remind you of a particular food?*

SCIENCE FAIR IDEA

What other solutions can make milk curdle? Try out different fruit juices or other edible liquids that you find in your kitchen and test how your milk reacts to these. Remember that acidity (how sour something is) and enzymes are good ways to make milk curdle!

SCIENCE FAIR IDEA

How much pineapple or lemon juice is necessary to make milk curdle? To find out, you could repeat this activity, but this time change the amount of juice that you add to your milk. *Is one drop enough or do you need 10? How many drops does it take to make the milk start to curdle?*

SCIENCE FAIR IDEA

In this activity, you tested one kind of milk. *Do you think other milks such as fat-free and lactose-free milk or coconut and almond milk will give you the same or similar results?* There is only one way to find out: test it!

OBSERVATIONS AND RESULTS

Did you see some nice clump formation in your milk? Whereas regular milk looks smooth and white, it changes very fast when you add a teaspoon of lemon juice. It almost immediately gets thicker in consistency, and you see white clumps forming that stick to the cup wall when you swirl the milk slightly. The clumps, or curd, consist of casein proteins that are usually in solution where they form micelle structures. These structures are very fragile, and when you change the conditions of the solution, they can easily break up and form clumps of casein proteins. This can happen if you change the pH, or acidity, of the milk, which means making it sourer. Lemon juice is very acidic, and that is why you see the casein proteins clumping once you add it. Heating the lemon juice does not affect its acidity, which means when you added heated lemon juice to your milk, the exact same reaction should have occurred.

CLEANUP

You can throw away the curds and liquids from your experiment. Wash the materials with soapy water and return everything to the place where you found it.

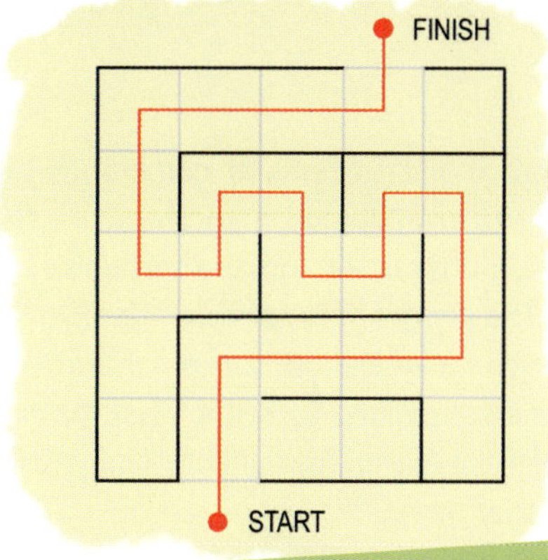

Program with Paper

A-Maze Your Friends with a Handwritten Computer Program

LEARN MORE ABOUT HOW COMPUTER PROGRAMS WORK BY WRITING ONE YOURSELF—TO SOLVE A MAZE OF YOUR OWN CREATION!

You probably use computer programs every day. Every time you are on the internet, play a video game or use a smartphone you rely on computer programs. Do you know you can learn a little bit about programming without needing to use a computer? In this fun activity, you will write a "program"—a set directions for a volunteer to find the way through a maze. Can you help them find the way through without crashing?

PROJECT TIME

20 minutes

KEY CONCEPTS

Computers
Computer programming
Problem solving

BACKGROUND

A computer program is a list of instructions or commands that tell a computer what to do. For example, when you are typing on a computer, there are commands that instruct it to display certain characters on the screen when you push a certain key on your keyboard. The program for a video game might have different instructions that tell the game to move the on-screen character when you push on a controller. A web browser has instructions to tell it what to do when you mouse click on different buttons. There are many, many more examples of what computer programs can do. Can you think of more?

Computers are very good at following directions—but that can lead to a problem: What happens if a human programmer accidentally gives a computer the wrong directions? The computer can't think for itself—it will go ahead and follow the directions it was given, even if they're wrong! This can cause the program to "crash," or not function as it was intended to. Computer programmers call errors in computer code "bugs" and the process of finding and removing them is known as "debugging." In this project, you will write a set of directions for a volunteer to navigate a maze, using commands like "move forward," "turn left," and "turn right." If you make a mistake with your directions, the volunteer will "crash" into the wall of the maze, and you will have to debug your program and try again!

MATERIALS

- Graph paper
- Ruler
- Pencil
- Volunteer

PREPARATION

- Use a sheet of graph paper to create a small maze that fits on a rectangular grid. All of the maze's walls should be vertical or horizontal—no diagonal or curved lines.
- Start out with a small, easy-to-solve maze. (You can make a more difficult one later.)
- Make sure the maze has clearly labeled "start" and "finish" points.

PROCEDURE

- Imagine that you are a person walking through your maze, starting at the "start" point. On a separate piece of paper write directions, one step at a time, for how you would walk through the maze. For example, use commands like "move forward two squares," "turn left," and "turn right." Continue writing directions until you have enough to reach the maze's "finish" mark.

- **Optional:** If you prefer, you can draw arrows for your directional instructions instead of writing out the words. *Does it take a while to write out the step-by-step instructions? How does this process compare with drawing the maze?*

- Now, hand your maze, the directions, and a pencil over to your volunteer. Instruct them to follow your directions exactly as they go through the maze, even if they see a mistake. They are not allowed to solve the maze on their own or change the directions.

- Watch closely to see if the volunteer successfully completes the maze by following your directions. *Did they crash into any of the walls or make it all the way to the finish line?*

- If your volunteer made it all the way to the finish line, then congratulations! Your program did not have any bugs. Try making a bigger, more complicated maze and start over.

- If your volunteer crashed into one of the maze's walls, however, you need to debug your program! Carefully trace through your program step by step to see where things went wrong. *Did you tell them to move one block too far or turn one block too early?* Just one mistake can throw off your entire program and cause a crash later on, so look carefully!

- After you discover the bug, rewrite your program. If you only need to change one or two steps, you can just erase them instead of rewriting the whole thing.

- Have your volunteer try the maze again, following your new program. *Did they solve it this time?* If not, keep debugging your program until they solve the maze. *How difficult was the process of finding and fixing bugs for your maze instructions?*

EXTRA

Computer programs are written in many different "languages," which all have slightly different formats but very similar underlying concepts. For example, some computer programs use punctuation and spelling differently—this is called the program's "syntax." *Can you come up with a different "language" that accomplishes the same task of solving the maze? For example, what about using commands that are defined relative to absolute directions on the piece of paper like "move up," "move down," "move left," and "move right;" as opposed to turning left or right as if you were a person standing in the maze? Although these two languages are different, can you see how they both be used to do the same thing?*

EXTRA

Try this project with more complicated mazes that include diagonal or curvy lines. *If you can no longer rely on a rectangular grid for movement, how would you write the program?* What sort of language might you need to introduce? (Hint: You can also try introducing new commands like "move forward until you reach an intersection" or "move forward until you hit a wall.")

EXTRA

Instead of using a maze on paper, write a program to navigate a person walking through your house or school. *Can they move from room to room without bumping into a wall by only following your directions?*

OBSERVATIONS AND RESULTS

You probably found that it only takes one bug early on in a program to cause a crash. When solving a maze, there is little room for error—make one wrong turn and you can—literally—crash by bumping into a wall or you can get lost!

Have you ever used a computer program that crashed or played a video program with a glitch? Did you find the experience frustrating? This is why companies do lots of debugging and testing before they release a program to the public—and even so, sometimes they have to release "patches" or "updates" for a program after a bug is discovered.

CLEANUP

Recycle any used paper. Put away the other materials you used.

THE SCIENTIFIC METHOD

The scientific method helps scientists—and students—gather facts to prove whether an idea is true. Using this method, scientists come up with ideas and then test those ideas by observing facts and drawing conclusions. You can use the scientific method to develop and test your own ideas!

Question: What do you want to learn? What problem needs to be solved? Be as specific as possible.

Research: Learn more about your topic and refine your question.

Hypothesis: Form an educated guess about what you think will answer your question. This allows you to make a prediction you can test.

Experiment: Create a test to learn if your hypothesis is correct. Limit the number of variables, or elements of the experiment that could change.

Analysis: Record your observations about the progress and results of your experiment. Then analyze your data to understand what it means.

Conclusion: Review all your data. Did the results of the experiment match the prediction? If so, your hypothesis was correct. If not, your hypothesis may need to be changed.

GLOSSARY

contamination: Being made impure by polluting.
dilute: To add water to something.
emulsion: A mixture of two or more liquids that are normally unable to be mixed together.
enzyme: A chemical in plants and animal bodies that aids natural life processes, such as digestion.
faint: Barely able to be seen.
gelatinous: Having the properties of gelatin or jelly, such as being wiggly or squishy.
generate: Create.
hue: A pure pigment or color.
mimic: To imitate closely.
perceive: To be aware of something.
perforated: Having holes.
pigment: Matter that gives color to something.
proportionate: Corresponding in size or intensity.
quantify: To measure the amount or quantity of something.
soluble: Able to be dissolved in liquid.
yield: To produce.

ADDITIONAL RESOURCES

Books

Dickinson, Michelle. *Kitchen Cabinet Science Projects: Fifty Amazing Science Experiments to Make with Everyday Ingredients.* New York, NY: Penguin Workshop, 2021.

Heinecke, Liz Lee. *Sheet Pan Science: 25 Fun, Simple Science Experiments for the Kitchen Table: Super Easy Setup and Cleanup.* Beverly, MA: Quarry Books, 2022.

Holiday Science: 10 Fun Chemistry and Food Science Projects. New York, NY: Scientific American Educational Publishing, an imprint of Rosen Publishing, 2023.

Websites

Discovery Education
Sciencefaircentral.com

Exploratorium
www.exploratorium.edu/explore/activities

Science Buddies
www.sciencebuddies.org/science-fair-projects/project-ideas/list

Science Fun for Everyone!
www.sciencefun.org/?s=science+fair

Videos

Air Pressure | DIY Science Time
https://wtvi.pbslearningmedia.org/resource/82ce7518-080f-4536-b114-8b78089b7054/air-pressure-video-diy-science-time/

Balloon Science | DIY Science Time
https://wtvi.pbslearningmedia.org/resource/f5084e86-a375-4158-b89f-bc4fbd25c9b2/diy-science-time/

Publisher's note to educators and parents: Our editors have carefully reviewed these websites to ensure that they are suitable for students. Many websites change frequently, however, and we cannot guarantee that a site's future contents will continue to meet our high standards of quality and educational value. Be advised that students should be closely supervised whenever they access the internet.

INDEX